On Matter, Mass, and Motion

Greg Feild

September 14, 2017

Thus scientific knowledge is a demonstrative state, . . .

i.e. a person has scientific knowledge when his belief is conditioned in a certain way, and the first principles are known to him; because if they are not better known to him than the conclusion drawn from them he will have knowledge only incidentally.

This may serve as a description of scientific knowledge.

-- Aristotle,
Ethics

About the author:

Greg Feild is a physicist.

He has a PhD and everything!

Abstract:

In the universal model of our sinister universe,
nature (i.e. all interaction) is mechanical
and deterministic.

In this book, we explore this strange new world!

In addition, we offer the usual corrections and clarifications
concerning parts of our model still under development,
and summarize the current status of this totally awesome,
completely mind-straightening, universal model of the world.

let's do physics !

G_F

Errata:

To date, we have been ~~insistent~~ positing that the electron neutrino has a magnetic moment, mu_v, and this magnetic moment is proportional to the mass of the neutrino, m_v, the electric charge, e, and an incorrect factor of 1/c for some reason ... ;

$$mu\_v = e*hbar/2*m\_v*c \tag{a}$$

We still expect the neutrino to have a magnetic moment, but the factor of e is not compatible with recent developments in our theory, nor is it 'parallel' with our expression for the gravitational contribution to the generalized Lorentz force; a generalization which was motivated by our assuming a magnetic moment for the spinning mass of the neutrino!

The 'logical' solution is to replace the coupling constant, e, with the square root of the gravitational constant G divided by 4*pi*epsilon_0, and multiply by the relativistic mass (and remove the factor of 1/c).

$$mu\_v = (G/4\pi\varepsilon)^{½}(m)(hbar/2*m\_v) \tag{b}$$

We will explore how we arrived at this formula later in this paper.

The mass of the electron is still presumed to be

$$m\_e = e*m\_v \tag{c}$$

(where e is the *magnitude* of the electric charge), but the electron magnetic moment would now be given by

$$mu\_e = e*hbar/2*m\_e + (G/4\pi\varepsilon)^{½}(m)(hbar/2*m\_e) \tag{d}$$

Speaking of the Lorentz force, we've encountered a lot of trouble (made careless errors) in crafting our recommended generalization. Here is the latest offering.

The complete, 'classical', relativistic, Lorentz force between two identical electrons is (now)

$$\mathbf{F} = (G/c^2 - (e/m_e)^2(\mu/4\pi))(c/R)^2(m_1 m_2 \mathbf{r} + (1/c^2)(\mathbf{p_1} \times \mathbf{p_2} \times \mathbf{r})) \tag{e}$$

Finally, what we have been calling the 'matchbook summary', now looks like this;

$$L\_interaction = -i*hbar(G^{1/2} - e/m_{rest})(\psi^{bar} \gamma^{\mu} A_{\mu} \partial\psi/\partial t) \tag{f}$$

Now, everything should be parallel, even if not fully correct!

Preface:

Intrinsic angular momentum, magnetic moments, spin, polarization, helicity ....

Confounding and confusing!

And, the foundation of our physical world; the fundamental "expression" of energy and momentum; the very essence of matter.

That's all.

The world is all about spin. Planets, stars, galaxies, gas molecules, electrons, neutrinos, photons; all have 'intrinsic' angular momentum, or spin; a.k.a. *absolute motion*. What things are missing from this list? Balls, buckets, tops, gyroscopes, merry-go-rounds, etc.; all human sized objects. Familiar everyday objects do not spin unless one applies an external torque and, of course, they all will eventually stop rotating due to friction. We hypothesize this is why the importance of the role of spin in both microscopic and cosmological interactions has been generally minimized and overlooked in the past.

I have personally been quite flummoxed by magnetic moments; not only due to an incomplete understanding of the standard model representation, but also in reconciling this representation with our theory that magnetic moments are due to spinning particle mass.

In this current paper, we will reexamine the magnetic moment; salvaging (?) our theory and reconciling it with the current standard model theory, which, of course, is brilliant and correct!

Actually, we will derive the 'relativistic' magnetic moment for the electron, which reduces to the standard model representation as in the first term of equation (d), for an electron at rest.

On the other hand, our notion of the conservation of the "electromagnetic charge" does not seem quite so easy to salvage; or at least not worth the effort at the moment.

(This theory joined the model late in the game, and it really did not contribute very much!)

Our theory of 'weak isospin' should still be strong.

Big-ly!

Introduction:

Where to begin?

We live in a world of curved space, parallel universes, and collapsing wave functions; where every person spawns a million branching realities with every breath they take.

The atomic theory of matter has been completely abandoned in favor of a myriad complex of waves and fields.

It seems any problem can be solved with more ~~epicycles~~ fields!

So much for the working hypothesis that nature is simple and elegant in design.

Why is curved spacetime supposed to be more appealing than 'action at a distance'? Are they not, essentially, equally 'occult'?

No matter. It is a giant leap from saying 'in our model we imagine spacetime to be curved' to declaring emphatically that 'spacetime is curved', or that time is not real, nor motion, etc.

These ideas fly in the face of common sense, and yet people hold them fiercely.

Even the physicist who believes that time and motion are nothing but an illusion will jump out of the way of a speeding bus; one would hope!

We propose any physical theory must pass "The Speeding Bus Test".

Nature is mechanical and *dynamical*. Particles and their interactions can be completely described and explained in terms of fundamental, 'solid', and *real* units of matter, constantly in motion, and continually exchanging energy and momentum.

That is the premise of the universal model.

The Lorentz force:

Let's begin with our usual visit to the Lorentz force; a veritable new gold mine of information!

In this section, we will review what we have learned to date, and then see what other information we might extract from this equation, and what further generalizations we may make about the nature of forces and particle interactions.

The complete, 'classical', relativistic, Lorentz force between two identical electrons is

$$\mathbf{F} = (G/c^2 - (e/m_e)^2(\mu/4\pi))(c/R)^2(m_1 m_2 \mathbf{r} + (1/c^2)(\mathbf{p}_1 \times \mathbf{p}_2 \times \mathbf{r})) \qquad (1)$$

where, of course, $\mathbf{F}_1 = -\mathbf{F}_2$. Newton's second law tells us

$$\mathbf{F}_1 = m_1 \mathbf{a}_1 \qquad (2)$$

If we compare equations (1) and (2) we can see that the acceleration of a particle is *independent of its mass*.

$$\mathbf{a}_1 = \text{Function}(m_2, \mathbf{R}) \qquad (3)$$

This is a general result that we used to assume applied only to the gravitational interaction.

In addition, the relative strengths of the four terms in equation (1), or the 'four forces of classical physics', are approximately as follows (superseding previous estimates ...);

electricity = 1 ; magnetism ~ $1/c^2$; gravity ~ G/c^2 ; magnetic gravity ~ G/c^4

I think Maxwell would approve!; except that we have no more need for his fields.

The two electrons exert equal and opposite forces on one another during the interaction, and the evolution of the force is completely described by the (variable) mass-energy of the two electrons, $m_1(\mathbf{r}_1(t))$, $m_2(\mathbf{r}_2(t))$, where the time, t, is *common* to both electrons.

Of course, without the fields there is no mathematical or physical mechanism to explain the interaction of these two particles 'at a distance'.

We like the idea of 'one virtual photon' (which, of course, is a discovery of the field theory model!) constantly coupling the particles; a time varying conduit for energy and momentum exchange. Can we infer or derive the virtual photon without resort to field theory?

We will defer this investigation for now. A challenge to the reader, perhaps.

Let's look at our 'new' Lorentz force in a little more detail. If we define

$$K == (G/c^2 - (e/m_e)^2(\mu/4\pi)) \qquad (4)$$

then we can write equation (1) as

$$\mathbf{F} = K*(c/R)^2(m_1 m_2 \mathbf{r} + (1/c^2)(\mathbf{p_1 x p_2 x r})) \qquad (5)$$

Next, we factor out the particle masses from the momentum term

$$\mathbf{F} = K*(c/R)^2(m_1 m_2 \mathbf{r} + (1/c^2)(m_1 m_2 \mathbf{v_1 x v_2 x r})) \qquad (6)$$

$$\mathbf{F} = K*(c/R)^2(m_1 m_2)(\mathbf{r} + (1/c^2)(\mathbf{v_1 x v_2 x r})) \qquad (7)$$

where **r** is the unit vector **R**/R.

Since our two masses form a closed, conservative system, we can 'normalize' our force by dividing by the total energy of the system; $E_{TOT} = m_1 + m_2$.

$$\mathbf{F}/E_{TOT} = K*(c/R)^2 \mu (\mathbf{r} + (1/c^2)(\mathbf{v_1 x v_2 x r})) \qquad (8)$$

where $\mu(\mathbf{R}, d\mathbf{R}/dt) = m_1 m_2/(m_1 + m_2)$ is the reduced mass of the two body system.

In order to investigate the vector cross product term, we will assume our two masses (no longer necessarily electrons) are equal and orbiting one another.

Then we can write

$$\mathbf{F}/E_{TOT} = K*(c/R)^2 \mu (1 - (1/c^2)(v^2)) \qquad (9)$$

$$\mathbf{F}/E_{TOT} = K*(c/R)^2 \mu - K*(\mu v^2/R^2) \qquad (10)$$

We will call the second term in equation (10), the *coriolis* force, because ...

Why not?   :)

If we recast our force equation into polar coordinates and allow $m_1 \neq m_2$ (i.e. for the study of planetary motion; Kepler's equations), we will pick up the usual *centrifugal* force term, in addition to our new *coriolis* force term.

Finally, there will be a force term corresponding to the interaction of the spin/angular momentum (σ, $\mathbf{l}$) of one object with the 'magnetic force vector' of the other object; F_{spin}. We defer a further discussion of this idea for now.

(Remember, in our model, the relativistic mass of an object is due to the *total* relativistic motion of the mass; including spin!)

So, the total Lorentz force, for cosmology for example, will consist of four terms;

$$F_{universal} \sim F_{central} + F_{centrifugal} + F_{coriolis} + F_{spin} \qquad (11)$$

Inertial reference frames:

Because spin is an inherent component of our theory, and because everything is spinning, there can be no inertial reference frames, even in principle!

We suggest the inertial observer (who is always 'at rest') reference their inertial coordinate system to the " 'fixed background' of 'empty space' ". (You may rearrange the "scare quotes" as you'd like!)

The 'fixed stars' are no longer fixed, nor must we worry about their influencing our measurements. The stars will either be part of our study, or too far away to matter.

As to spinning buckets, we still make no hypothesis.

Life needs some mystery!

:)

The Dirac equation:

The Dirac equation for a free electron is

$$H \psi = (\alpha \cdot p + \beta m_0) \psi \quad (12)$$

Our 'gauge invariant' solution as intimated in "A critical examination of classical and quantum mechanical waves" would look something like this;

$$\psi = \exp(-i^*m_0) \exp(-iE^*t) \exp(i\mathbf{p} \cdot \mathbf{x}) \quad (13)$$

Replacing E(t) with the relativistic mass m(t), gives

$$\psi = \exp(-i^*m_0) \exp(-im^*t) \exp(i\mathbf{p} \cdot \mathbf{x}) \quad (14)$$

The first factor in equation (13) is a global phase factor representing the invariant rest mass-energy/charge of the particle. This is an extra and annoying charge which is already included in the second factor of the equation; i.e the total mass-energy/charge, m(t), which we take to be our locally variable 'phase factor' *and* the locally variable interaction charge of the particle.

However, our theory is not really a gauge theory, and there seems to no formal way (e.g. by a series expansion) to have the global phase factor in equation (14) cancel the rest mass energy term in the Dirac equation.

So ... we take our lead from the standard model and throw the rest mass term away!

However, we needn't 'recover' the rest mass from an interaction with the Higgs field as in the standard model. In our model the contribution of the rest mass energy is already accounted for in the total relativistic mass energy of the particle, which is the particle's interaction charge.

To introduce interactions into the Dirac equation, we modify the standard replacement

$$p_\mu \rightarrow p_\mu + eA_\mu \quad (15)$$

to include the gravitational interaction, and to account for the relativistic mass as the fundamental interaction charge, and obtain

$$p_\mu \rightarrow p_\mu - i^*\hbar (G^{1/2} - e/m_{rest})A_\mu \, \partial/\partial t \quad (16)$$

This substitution should then yield the Universal Interaction Lagrangian of equation (f).

$$L\_interaction = -i\hbar(G^{1/2} - e/m_{rest})(\psi^{bar}\gamma^{\mu}A_{\mu}\partial\psi/\partial t) \qquad (17)$$

We interpret the component A_{μ} as representing a *virtual* photon, and thus we have no need for the antisymmetric tensor $F^{\mu\nu}$, and the corresponding E and B fields, to facilitate the interaction or to carry energy and momentum. Remember, in our model, all the energy and momentum in an interaction is carried by the particles!

The virtual photon is already coupled to, or 'tethered' between, two mass-charge currents, only one of which is indicated in equation (17). The second current could be a similar leptonic current or a real photon current!

In our model, we expect to treat the real and virtual photons as separate particles, each with their own 'wave function'.

Gauge theory:

Technically, our theory is *not* a gauge theory, although it does exploit and explain the observed gauge invariance of the electromagnetic interaction (another ingenuous discovery of the standard model, to be sure!).

In our model, the resolution of the 'gauge invariance issue' does not involve, or allow, the capricious variation of local charge with corresponding and compensating potential fields. (This is a feature never truly exploited, or properly explained, or *necessary* in QED ....)

Instead, in the universal model, we have a locally varying charge because the mass-energy is the coupling charge of a particle, and this varies during an interaction.

In conclusion, since the standard model explanation (and implementation) of the required 'gauge invariance' observed in QED is now thought to be incorrect, the generalization of gauge theory to explain the weak and strong interactions (thus yielding the W, the Z, and eight gluons) is probably also incorrect.

Of course, we have already removed the W, the Z, and gluons and quarks from our model.

However, the more corroborating arguments for a new theory, the better.

The quantum of action:

Our rather disdainful dismissal of the quantum of action (in "On Wave Particle Duality and the Quantum of Action", no less!), as a glorified conversion constant was both hasty and ill-advised, particularly considering our theory of the mechanical nature of the photon as an harmonically oscillating polarization, or angular momentum vector traveling at the speed of light!

What we *meant* to say was that the quantum of action, h, is the fundamental unit of matter and is inextricable, and inseparable, from particle spin.

The photon:

The photon is one "free" quantum of action. Our photon is an inertialess, massive particle of 'spin = 1' (i.e. the photon has an angular momentum, L = h). The angular momentum vector is aligned along the direction of motion of the photon. This projection of the angular momentum L along the direction of motion (the photon polarization), oscillates harmonically with a frequency, ν = E/h.

A 'plane polarized' photon switches from left handed polarization to right handed polarization every 2π radians. This 'flipping' constitutes the energy and linear momentum of the photon. The angular momentum of the photon is always equal to h.

This model is to be contrasted with the *incorrect* theory of the photon as an inertialess blob of energy that somehow "swells up" in an undefined way as it acquires more energy and momentum.

Imagine a standing, plane polarized, electromagnetic wave. The electric field oscillates harmonically exhibiting the usual characteristic nodes where the field strength goes to zero.

This standing wave is made up of photons. How do the photons manifest as an oscillating electric field?

The photon polarization behaves like a simple harmonic oscillator. Just like our particle in a box, it cannot interact as the angular momentum vector passes through the zero point. Also, when the photon is spinning in one direction, the associated electric field vector points up, when it is spinning in the other direction, the electric field vector points down.

This idea is worth repeating. A free photon cannot physically interact when the photon polarization is passing through zero. This model of the photon is able to explain the partial transmission and partial reflection of light waves incident on on a thin sheet of glass.

The neutrino:

The neutrino is one "bound" quantum of action. The neutrino has an intrinsic angular momentum of L = hbar/2. We note that

$$\text{hbar}/2 = h/4\pi \tag{18}$$

where the factor of 4π is interpreted to be the 'solid angle integral'.

Hence, we conclude the electron neutrino is one unit of intrinsic angular momentum, h, per unit volume of space.

This is all we have to say about the neutrino for now.

In our next paper, we shall try to use this observation to derive the neutrino mass, unless a 'challenged reader' beats us to the punch!

The running of alpha:

In the standard model, the electromagnetic coupling strength is expressed in terms of the electric charge, e;

$$\alpha = e^2/4\pi\varepsilon\,\text{hbar}\,c \tag{19}$$

In our model, the 'electric charge' looks like this

$$e \rightarrow (m^*e/m_e) \tag{20}$$

and alpha becomes

$$\alpha = m^2(e/m_e)^2/4\pi\varepsilon\,\text{hbar}\,c \tag{21}$$

$$\alpha = (m_e^2/1 - v^2/c^2)(e/m_e)^2/4\pi\varepsilon\,\text{hbar}\,c \tag{22}$$

$$\alpha = (1/1 - v^2/c^2)\,e^2/4\pi\varepsilon\,\text{hbar}\,c \tag{23}$$

$$\alpha = \alpha_0(1 + (v/c)^2 + (v/c)^4 + \ldots) \tag{24}$$

In our running of alpha, the velocity squared replaces the four-momentum transfer, Q^2, and there is no need to introduce an arbitrary cut-off mass.

Magnetic moments:

The standard model formula for the magnetic moment of the electron is

$$\mu_e = e\hbar/2 m_e \qquad (25)$$

where, it seems, m_e is the *rest mass* of the electron. This formula can be 'derived' from some fairly dubious hand waving arguments, and, of course, it "falls out naturally" from the Dirac equation.

In a previous paper, we were curious about the behavior of the magnetic moment at relativistic speeds. We naively replaced the rest mass, m_e, with the relativistic mass in equation (25). The results did not make sense (although we tried to 'make hay' with them anyway!)

The electron magnetic moment decreased at relativistic speeds. This seemed to run counter to common sense, *and* to our theory that the electron magnetic moment arises from spinning mass, and our theory that the electron "spins faster" when it is accelerated.

Since our electromagnetic coupling charge is $e \rightarrow (m^* e/m_e)$, we will make the same substitution as we made in our study of the running of alpha.

$$\mu_e = (m/m_e)(e\hbar/2 m_e) \qquad (26)$$

Similarly, in our model, the gravitational coupling charge is considered to be $G^{1/2} m$, so our naive guess for the neutrino magnetic moment would now be

$$\mu_\nu = G^{1/2}(m)(\hbar/2 m_\nu) \qquad (27)$$

Unfortunately, this formula is not dimensionally correct, so we add a factor of $4\pi\varepsilon$ 'by hand'.

$$\mu_\nu = (G/4\pi\varepsilon)^{1/2}(m)(\hbar/2 m_\nu) \qquad (28)$$

It looks like this part of the model still needs some work.

A challenge to the reader!   :)

Conclusion:

Our new model is now 'complete'.

I hope the physics community will adopt it, formalize everything,
and whip it into proper shape.

I reckon the results should also be compared with existing data, just as a check!

As far as new experiments are concerned, the structure function results from the Jefferson Lab collaborations seem quite intriguing. These measurements should be examined in light of our new theory of the proton as a bound state of two positrons and an electron.

Elsewhere, the world over, people are devising new neutrino experiments.

These experiments are exactly what we need just now to confront the basic premise(s) of new our model.

Serendipity!

Spooky.

Summary:

Here's a 'back of the envelope summary' of the universal model (to date!), including results from several of our previous papers.

The Lorentz force:

$$\mathbf{F} = (G/c^2 - (e/m_e)^2(\mu/4\pi))(c/R)^2(m_1 m_2 \mathbf{r} + (1/c^2)(\mathbf{p_1} \times \mathbf{p_2} \times \mathbf{r}))$$

The QED Lagrangian:

$$L\_interaction = -i*hbar(G^{1/2} - e/m_{rest})(\psi^{bar} \gamma^{\mu} A_{\mu} \partial\psi/\partial t)$$

The running of alpha:

$$\alpha = \alpha_0(1 + (v/c)^2 + (v/c)^4 + \dots)$$

alpha_G:

$$alpha\_G = m_e^2 * G/(hbar*c) \quad ; \; m_e \text{ is the relativistic mass.}$$

alpha_weak:

$$alpha\_weak = m_v^2 * G/hbar*c \quad ; \; m_v \text{ is the relativistic mass.}$$

alpha_strong:

Please consult "The Sinister Universe".

The nucleons:

$$p = (e+, e+, e-) \; ; \; n = (e+, e-, \nu_e bar)$$

The pion:

$$\pi^0 = 1/(2)^{1/2} (e\, e\hat{}bar - mu\, mu\hat{}bar)$$

The Higgs boson:

$$H = (nu\_e, nu\_e\hat{}bar)$$

The leptonic table:

LEPTONS   ANTI-LEPTONS

electron	electron neutrino	PARITY ⇔	electron antineutrino	positron
⇐	CHARGE	MASS ⇵	CHARGE	⇒
muon	muon neutrino	PARITY ⇔	muon antineutrino	anti-muon
⇐	CHARGE	MASS ⇵	CHARGE	⇒
tau	tau neutrino	PARITY ⇔	tau antineutrino	anti-tau
⇔	weak isospin	mass isospin ⇕	weak isospin	⇔

TABLE 1: The leptons and their interrelations; or the kleptogenesis of the leptoquarks.

Any lepton can be 'generated' from any other by the appropriate applications of the parity operator, the weak isospin operator, and our newly proposed 'mass isospin' operator.

Muon decay:

The muon "sheds" a muon neutrino to become a "generic" virtual charged lepton.

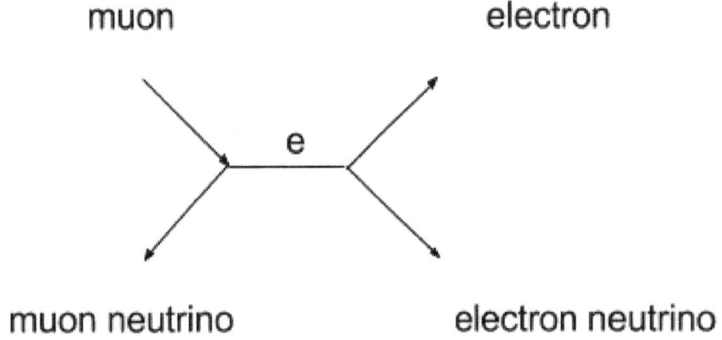

FIGURE 1: Muon decay. The propagator is most likely a generic virtual lepton, e/mu/tau.

References:

Modern Elementary Particle Physics
Gordon Kane

Classical Dynamics of Particles and Systems
Jerry B. Marion

Foundations of Electromagnetic Theory
John R. Reitz, Frederick J. Milford, Robert W. Christy

Quantum Physics
Rolf G. Winter

Gauge Theories in Particle Physics
I. J. R. Aitchison and A. J. G. Hey

Quarks and Leptons: An Introductory Course in Modern Particle Physics
Francis Halzen, Alan D. Martin

Quantum Field Theory
F. Mandl, G. Shaw

Theoretical Mechanics of Particles and Continua
Alexander L. Fetter, John Dirk Walecka

and

Elementary Modern Physics   (Best Book Ever!)
Richard T. Weidner, Robert L. Sells

Books by Greg Feild:

1. "A quantum mechanical theory of gravitational interactions"
   CreateSpace Independent Publishing, 8/29/2016

2. "Observations on the quantum mechanical nature of gravity"
   CreateSpace Independent Publishing, 10/8/2016

3. "On gravitation and electric charge"
   CreateSpace Independent Publishing, 10/29/2016

4. "On spin, mass, and charge"
   CreateSpace Independent Publishing, 11/29/2016

5. "On angular momentum, acceleration, and absolute motion"
   CreateSpace Independent Publishing, 1/1/2017

6. "The Sinister Universe"
   CreateSpace Independent Publishing, 3/1/2017

7. "On Parity and Isospin"
   CreateSpace Independent Publishing, 4/11/2017

8. "Reflections on the Sinister Universe"
   CreateSpace Independent Publishing, 5/12/2017

9. "On Current Physics"
   CreateSpace Independent Publishing, 6/11/2017

10. "A Critical Examination of Classical and Quantum Mechanical Waves"
    CreateSpace Independent Publishing, 6/18/2017

11. "On wave particle duality and the quantum of action"
    CreateSpace Independent Publishing, 7/6/2017

Compilation:

"The Universal Model of Our Sinister Universe: The First Ten Books"
CreateSpace Independent Publishing, 7/2/2017

Notes: :)

Secret song:

    Or … we could call it the **SUM** (total of everything!) model.

        The **S**inister **U**niverse **M**odel

    Or … we could just call it greg's model.

        Grass **R**ooted, **E**mergent **G**ravity and **S**pace

        grass root physics!

www.ingramcontent.com/pod-product-compliance
Lightning Source LLC
Chambersburg PA
CBHW082225220526
45470CB00010B/3316

9 781976 350283